THE FUTURE OF YOUTUBE

THE FUTURE OF YOUTUBE

ELARA PHOENIX

CONTENTS

1 Introduction 1
2 Increasing User Engagement 5
3 Emerging Video Formats 9
4 Monetization Strategies 13
5 Influencer Marketing 17
6 Artificial Intelligence Integration 21
7 User-Generated Content 25
8 International Expansion 29
9 Live Streaming 31
10 Virtual Reality and Augmented Reality 35
11 Data Analytics and Personalization 37
12 Copyright and Content Regulation 41
13 Advertising and Brand Partnerships 45
14 Mobile Optimization 49
15 Social Media Integration 53
16 Challenges and Opportunities Ahead 57

Copyright © 2024 by Elara Phoenix
All rights reserved. No part of this book may be reproduced in any manner whatsoever without written permission except in the case of brief quotations embodied in critical articles and reviews.
First Printing, 2024

CHAPTER 1

Introduction

In the rapidly evolving digital ecosystem, it is crucial to look beyond immediate data and consider long-term trends that will shape the future of content creation and consumption. This study aims to help stakeholders evaluate the merits of different strategies and predict their potential outcomes. By building on leading indicators of YouTube's short-term performance, we establish relationships between these indicators and longer-term metrics that are more difficult to observe. Specifically, we analyze video upload dynamics, audience engagement, and channel promotion features to create predictive models that anticipate video content popularity up to two years in the future.

Our core theme is inspired by the collaborative efforts of Community TV with the authors, aiming to bridge the gap between Social and Computer Sciences. By studying the behaviors of a complex social system influenced by individual actions, we strive to unify these perspectives and offer a comprehensive understanding of the digital landscape.

YouTube has experienced phenomenal growth, attracting an ever-increasing number of users and video uploads. Although not all video-sharing platforms witness such explosive growth, the general public is undoubtedly gravitating towards online media consumption. Today's viral videos command audiences that are ten or even a hundred times larger than those of a few years ago. As a result, content creators are adopting best practices and tailoring their content for online distribution. Recent trends suggest that this could be the pivotal year in which broadband video surpasses cable television in audience size in the U.S.

Context and Significance

Understanding these trends is essential for various stakeholders, including content creators, marketers, and platform developers. For content creators, insights into long-term popularity trends can inform content strategies, ensuring they remain relevant and engaging. Marketers can leverage these predictions to optimize advertising campaigns, targeting audiences more effectively. Platform developers can use this knowledge to enhance user experience, recommending content that aligns with users' evolving preferences.

Research Objectives

1. **Analyze Video Upload Dynamics**: Examine the patterns and frequencies of video uploads and their correlation with audience engagement.
2. **Assess Audience Engagement**: Study how viewers interact with video content over time, including likes, comments, shares, and watch time.
3. **Evaluate Channel Promotion Features**: Investigate the impact of various promotional strategies, such as collaborations, advertising, and algorithmic recommendations, on video popularity.
4. **Develop Predictive Models**: Create robust models that can forecast video popularity based on the analyzed data, offering a glimpse into the future of digital content consumption.

Implications and Applications

The insights derived from this research have broad implications. Content creators can tailor their production schedules and themes to align with predicted trends. Marketers can allocate resources more efficiently, optimizing their reach and impact. Platform developers can refine their recommendation algorithms, enhancing user satisfaction and retention.

As the digital landscape continues to evolve, understanding these dynamics becomes increasingly important. By bridging the gap between social and computer sciences, we aim to provide a holistic view of the factors driving YouTube's success and the broader implications for the future of online video content.

CHAPTER 2

Increasing User Engagement

Comparative Analysis with Facebook
When comparing YouTube with Facebook in terms of audiovisual material, it becomes clear that YouTube offers lower barriers to submission and a broader potential scope for video postings, particularly when discussing organic reach. Producing higher quality content on YouTube doesn't necessarily mean that creators can consistently produce and market new videos every week. However, YouTube's ranking algorithms take into account the number of supporters, engagement metrics such as reviews, likes, and comments, which play a crucial role in how videos are ranked in search results, highlighted on front pages, subscription feeds, and even in recommendations for future viewing through tools like Watkins Media Auto.

Enhancing User Interaction with Community Features
YouTube has been proactive in fostering user interaction by enhancing its community tab features. This tool encourages creators to upload posts, share pictures, links, and text with their followers, akin to features found on Instagram. By allowing users to share diverse content, from food photos to inspirational quotes, YouTube offers

a more immersive experience for followers. This type of engagement is instrumental in strengthening relationships and transforming casual viewers into dedicated fans who not only watch videos but also engage deeply with the creator's broader activities.

Best Practices for Engagement

To optimize user engagement, creators should consider the following strategies:

1. **Regular and Consistent Posting**: While not every creator can post weekly, maintaining a regular schedule helps keep the audience engaged and anticipating new content.
2. **High-Quality Content**: Focus on producing well-researched and polished videos that provide value to the audience. High-quality content is more likely to receive positive engagement metrics.
3. **Interactive Elements**: Utilize polls, Q&A sessions, and live streams to directly interact with the audience and foster a sense of community.
4. **Engagement with Comments**: Actively responding to comments and engaging with viewers' feedback can significantly enhance viewer loyalty and encourage more interactions.
5. **Cross-Platform Promotion**: Leverage other social media platforms like Instagram, Twitter, and Facebook to promote YouTube content and drive traffic across different channels.

Leveraging Analytics for Growth

Understanding and utilizing YouTube Analytics is vital for creators who wish to increase engagement. Key metrics to monitor include:

1. **Watch Time**: This measures how long viewers watch your videos, indicating content quality and audience retention.
2. **Audience Demographics**: Knowing who your audience is can help tailor content to their preferences and increase engagement.
3. **Engagement Rates**: Likes, shares, and comments provide insight into how well your content resonates with viewers.
4. **Traffic Sources**: Understanding where your viewers are coming from can help optimize promotional strategies.

Case Studies and Success Stories

Examining successful YouTube channels can provide valuable insights into effective engagement strategies. For instance, channels that focus on niche topics often have highly engaged audiences due to the specialized content they offer. Collaborations with other creators can also boost engagement by introducing new audiences to your channel.

Conclusion

Increasing user engagement on YouTube involves a multifaceted approach that includes high-quality content production, interactive community features, and strategic use of analytics. By adopting these practices, creators can build a loyal audience base, enhance viewer interactions, and ultimately achieve greater success on the platform.

CHAPTER 3

Emerging Video Formats

Live Streaming and Monetization

One of the less talked about but highly attractive monetization opportunities in the realm of live streams is the donations format. Although this may account for just 10% of a creator's income on average, for some top creators like PewDiePie, it can contribute up to 55-60% of their total revenue. Live streaming opens doors to alternative viewing formats that wouldn't fit well in traditional media or could thrive with an independent business model. Concerts, for instance, resonate particularly well with YouTube's live streaming capabilities. Twitch, primarily dedicated to live streaming e-Sports, showcases concerts successfully on YouTube. Midnight Oil's concerts, for example, attracted 800,000 and 240,000 viewers. These viewers not only watched the concerts live but often subscribed to the channel to enable chat and access other perks such as merchandise or opportunities to meet the creators in person.

Rise of Live Streaming

Live streaming has emerged as an increasingly popular trend. Between November 2015 and November 2016, the daily viewership of Facebook Live videos and YouTube live streams increased tenfold. This format offers significant advantages to creators. It's relatively easy to produce in terms of technology and investment, and easy to promote because platforms send out notifications that drive high click-through rates. Furthermore, real-time interaction with viewers enhances the likelihood of content going viral, bringing both visibility and monetization opportunities. Monetization can come in the form of virtual features such as 'Super Chat' and 'Super Stickers,' which YouTube introduced in 2017 in select countries and has since expanded to 66 countries across 43 languages. These features allow viewers to purchase chat messages or stickers that stand out during a live chat, helping them get noticed by the creator while also supporting the channel financially.

Advantages of Live Streaming

1. **Real-Time Interaction**: Live streaming allows creators to engage with their audience in real time, fostering a sense of community and immediacy.
2. **Ease of Production**: Compared to pre-recorded videos, live streams require less post-production effort and can be more spontaneous.
3. **Promotion and Notifications**: Platforms like YouTube and Facebook send notifications to subscribers when a live stream begins, resulting in higher engagement rates.
4. **Monetization Opportunities**: Features like 'Super Chat' and 'Super Stickers' provide additional revenue streams for creators.

Case Studies
PewDiePie

As one of the top creators on YouTube, PewDiePie has leveraged live streaming to significantly boost his revenue. Donations through live streams account for a substantial portion of his income, demonstrating the potential of this format for other creators.

Midnight Oil

The band Midnight Oil has successfully used live streaming to reach large audiences. Their concerts on YouTube not only attracted hundreds of thousands of viewers but also led to increased channel subscriptions and fan engagement through the chat feature.

Conclusion

Emerging video formats like live streaming offer creators new ways to engage with their audience and monetize their content. By enabling real-time interaction and offering innovative monetization tools, live streaming is set to become an increasingly important part of the digital content landscape. As more creators and viewers embrace this format, its impact on the media industry will continue to grow, offering exciting opportunities for both content creators and audiences alike.

CHAPTER 4

Monetization Strategies

Building Brand Reputation and Operational Efficiency
To build a strong brand reputation and enhance operational efficiency, it's essential to evolve your channel in a way that showcases your expertise. For example, if you're a gamer, consider demonstrating your superior skills by covering all levels of gameplay, showcasing mastery over different types of games, and segmenting your content into beginner guides versus endgame level content. This approach can help you attract a diverse audience and establish yourself as an authority in your niche.

Leveraging Expertise in Different Verticals
Once you've built up expertise in a specific space, you can leverage it in different verticals. For instance, you can earn revenue from affiliates by showcasing your gaming setup, offering beginner tutorials and advanced lessons, or hosting gamer community events. Planning events to cover future announcements in your niche can also yield media coverage opportunities and allow you to speak to different services to fund travel costs. This classic monetization technique enables you to reinvest profits to build clout in other areas, much like a typical media conglomerate. Once you've mastered audience development and content creation, you can apply these skills to other ventures.

Community Goals and Collective Success

Our community's ultimate goal is not only to increase the number of creators but to ensure that millions of independent businesses can operate at scale and thrive. Achieving this vision requires various strategies to bring sufficient revenue to creators, allowing them to invest in our collective creator pool and create content that resonates with audiences. Brand-building as a monetization strategy revolves around high-quality content. Building a collection of videos that can scale your brand will help you attract a monetizable audience, streamline operations through robust systems and processes, and develop valuable expertise that can create new revenue streams.

Key Monetization Strategies

1. **Brand-Building**: Focus on creating high-quality content that can scale your brand and attract a loyal audience. This involves developing a portfolio of videos that stand the test of time, making operations more efficient, and leveraging expertise to create new revenue streams.
2. **Direct Monetization**: Utilize direct monetization techniques such as ad revenue, sponsorships, and partnerships. This approach allows you to generate income directly from your content and collaborations with brands.
3. **Alternative Revenue Streams**: Explore alternative revenue streams such as merchandise sales, subscription services, and paid content. These options provide additional income sources and help diversify your revenue.

Tracking and Benchmarking

Tracking and benchmarking monetization strategies are crucial to understanding what works effectively. These strategies are constantly evolving, so it's important to stay updated on current trends and emerging techniques. By continuously highlighting what's working today and discussing emerging moves, creators can adapt and optimize their monetization approaches.

Conclusion

Building a successful and sustainable business as a content creator requires a multifaceted approach to monetization. By focusing on brand-building, direct monetization, and alternative revenue streams, creators can attract a loyal audience, generate sufficient revenue, and reinvest in their channels for long-term success. As the digital landscape continues to evolve, staying informed about effective strategies and adapting to new opportunities will be key to thriving in the ever-changing world of content creation.

CHAPTER 5

Influencer Marketing

YouTube: The Hub of Conversations

YouTube is a central hub where conversations happen. People turn to YouTube for advice, suggestions, and recommendations throughout their online purchasing journey. Collaborating with a charismatic personality who knows how to engage their audience is a natural and effective strategy for today's merchants. YouTube creators are adept at conveying messages. They invest their time in creating valuable, educational, and entertaining content, demonstrating their commitment to their audience and the material they present. This dedication makes YouTube an invaluable resource for marketers. YouTube creators excel at showcasing brands in a way that highlights their offers and product utilities.

The Rise of Influencer Marketing

Influencer marketing is set to become increasingly popular. While businesses today collaborate with influencers on platforms like Instagram and Facebook, YouTube influencer marketing is poised to reach new heights. According to reports, Google prioritizes YouTube video content in search results, enhancing the platform's impact. Videos themselves can drive conversions. People love "how-to" videos and product demos. When prospects watch a video explaining a product, they are 85% more likely to purchase it. This makes YouTube a prime venue for influencer marketing. In fact, 74% of consumers acknowledge that relationships and trust are the key reasons they choose an influencer.

Key Benefits of YouTube Influencer Marketing

1. **High Engagement**: YouTube videos often receive higher engagement compared to other platforms due to their visual and interactive nature.
2. **Trust and Authenticity**: Influencers on YouTube build strong relationships with their audience, fostering trust and authenticity that brands can leverage.
3. **Educational Content**: How-to videos and product demonstrations provide valuable information, helping consumers make informed purchasing decisions.
4. **SEO Benefits**: Google prioritizes video content from YouTube, making it easier for brands to improve their search rankings and visibility.

Successful Strategies for Influencer Marketing

1. **Identify the Right Influencers**: Look for influencers whose audience aligns with your target demographic. Their content should resonate with your brand values and goals.
2. **Develop Authentic Partnerships**: Collaborate with influencers in a way that feels natural and authentic. Forced partnerships can be easily spotted by audiences and may backfire.
3. **Create Engaging Content**: Work with influencers to create engaging and informative content that showcases your products in the best light.

4. **Leverage Analytics**: Use YouTube Analytics to track the performance of influencer campaigns, identifying what works and refining your strategies accordingly.

Case Studies
Example 1: Tech Gadgets
A tech company partnered with a popular YouTube tech reviewer to showcase their latest gadgets. The review videos included detailed demonstrations and honest opinions, which led to a significant increase in sales and brand awareness.

Example 2: Beauty Products
A beauty brand collaborated with several beauty influencers to launch a new product line. The influencers created tutorials and reviews, highlighting the products' features and benefits. This campaign resulted in high engagement and a boost in sales.

Future Trends in Influencer Marketing

1. **Live Streaming Collaborations**: Live streaming offers real-time interaction with audiences, providing an engaging platform for influencer collaborations.
2. **Micro-Influencers**: Collaborating with micro-influencers (influencers with smaller but highly engaged audiences) can be more cost-effective and yield high engagement.
3. **Long-Term Partnerships**: Building long-term relationships with influencers can create a more consistent and authentic brand presence.

Conclusion
YouTube influencer marketing is a powerful tool for brands looking to reach a wider audience and drive conversions. By leveraging the trust and authenticity of influencers, brands can create

meaningful connections with consumers. As the digital landscape evolves, staying ahead of trends and continuously refining influencer strategies will be key to sustained success in this dynamic field.

CHAPTER 6

Artificial Intelligence Integration

Understanding Audience Preferences

YouTube creators must grasp the kind of content that will resonate with their target audience before they even start creating it. This involves analyzing viewer preferences, trends, and engagement metrics to tailor content that meets audience expectations. Once the content is created, the next challenge is getting potential audiences to click on it. However, it is equally important to keep these audiences engaged once they start watching the videos.

Enhancing Engagement with Voice-First Apps

Witlingo has developed a platform that allows YouTube creators to engage their audiences using voice-first apps. This integration makes it easy for users to link their YouTube videos and playlists with their voice-first apps built on the Witlingo platform. These voice-first apps provide a new avenue for audience interaction, making the viewing experience more immersive and engaging.

Increasing Visibility with AI

AI can be utilized in numerous ways to ensure that creators' content gains more visibility in a cost-effective manner. AI tools can be used to enhance content titles and taglines, making them more appealing and search-friendly. Assistant apps can help YouTube creators optimize their content for search engines, ensuring better visibility in search results. Additionally, AI can assist in matching content with channels that have the most engaged audiences, ensuring that videos reach viewers who are more likely to interact and engage.

Practical Applications of AI on YouTube

1. **Content Optimization**: AI can analyze viewer data to provide insights on the types of content that are most likely to be successful. This includes recommendations on video length, format, and subject matter.
2. **SEO Optimization**: AI-driven tools can help creators optimize their content for search engines by suggesting relevant keywords, tags, and descriptions that improve search rankings.
3. **Audience Matching**: AI can help match content with audiences that are most likely to engage with it, enhancing viewer retention and interaction.
4. **Real-Time Analytics**: AI-powered analytics tools provide real-time insights into video performance, helping creators make data-driven decisions to improve their content strategy.
5. **Voice Integration**: Voice-first apps, powered by AI, allow for hands-free interaction with content, making it more accessible and engaging for users.

Future Trends in AI Integration

Over the next decade, various technologies will see different levels of adoption among YouTube creators. One of the most exciting is artificial intelligence, which can be applied in a variety of ways across the platform. Given that much of the content on YouTube is short (often less than a minute), keeping users engaged is critical. AI can help in this regard by enhancing various stages of content discovery and viewer interaction.

1. **Personalized Recommendations**: AI can provide personalized content recommendations based on viewer behavior and preferences, increasing the likelihood of engagement and retention.
2. **Automated Editing**: AI tools can assist in video editing, making the process faster and more efficient while maintaining high quality.
3. **Content Creation**: AI can be used to generate content ideas and even create basic video content, freeing up time for creators to focus on more complex projects.
4. **Enhanced Interaction**: AI-driven chatbots and voice assistants can facilitate real-time interaction with viewers, providing instant responses to queries and enhancing the overall viewing experience.

Conclusion

Integrating artificial intelligence into the YouTube ecosystem offers numerous benefits for content creators, from improving visibility and engagement to streamlining production processes. As AI technology continues to evolve, its applications on platforms like YouTube will expand, offering exciting new opportunities for creators to enhance their content and grow their audiences. Embracing these technologies will be key for creators looking to stay ahead in the competitive landscape of online video.

CHAPTER 7

User-Generated Content

Unique Characteristics of YouTube Content

YouTube user-generated content presents unique characteristics that set it apart from other types of internet text such as news articles or blogs. The site's rich media format significantly differentiates it, as it is composed of individual video "views" where creators record live events, share personal experiences, and disseminate self-generated knowledge. Unlike traditional documents like news articles or encyclopedias, these videos are often accompanied by sound, which adds an additional layer of complexity. This format, however, can pose accessibility challenges for individuals with visual and hearing impairments.

Consumption Patterns

The consumption of YouTube content differs from traditional media in several key ways. Videos frequently focus on non-fictional, real-life content aimed at informing or instructing viewers rather than creating emotional narratives through storytelling. As such, the purpose of user-generated videos often diverges from that of professionally generated content. Interestingly, YouTube and television viewership demonstrate an inverse relationship, suggesting that the platform is not in direct competition with traditional television programs.

The Impact of User-Generated Content

YouTube has fundamentally transformed how we access and view video content, enabling anyone to reach a global audience at any time. This capability has become a pervasive aspect of internet culture, often referred to as "user-generated content." Individuals produce videos for a multitude of reasons, such as sharing knowledge, expressing personal identities and experiences, and reacting to current events. The massive participation by users on the platform has made YouTube an essential and unique hub, leading to a wide range of political, social, and cultural impacts.

Socio-Economic and Cultural Implications

Throughout our report, we provide an in-depth characterization and analysis of the remarkable breadth of YouTube content. By capturing statistics that span all YouTube videos and times, we are uniquely positioned to consider the broader socio-economic and cultural implications of this content. YouTube has become a significant platform for political discourse, social movements, and cultural expression. This section will delve into the underlying features and challenges in this dynamic problem space.

Challenges and Opportunities

1. **Accessibility**: Improving accessibility for individuals with visual and hearing impairments is crucial. This includes better subtitles, audio descriptions, and other accessibility features.
2. **Content Moderation**: Ensuring that content adheres to community guidelines while balancing freedom of expression is a significant challenge.
3. **Monetization**: Creators must navigate the complexities of monetizing their content while maintaining authenticity and audience trust.

4. **Audience Engagement**: Understanding and leveraging audience engagement metrics is key to the success of user-generated content.
5. **Cultural Representation**: Ensuring diverse and accurate cultural representation in content is important for fostering a more inclusive platform.

Conclusion

User-generated content on YouTube represents a unique and vital component of the digital landscape. It offers an unparalleled platform for individuals to share their experiences, knowledge, and creativity with a global audience. As the platform continues to evolve, addressing the challenges and leveraging the opportunities will be essential for maximizing its positive socio-economic and cultural impacts.

CHAPTER 8

International Expansion

Mobile Unfriendliness and Global Mobile Browsing
While some videos are inherently mobile-unfriendly—needing larger display surfaces or relying on multiple simultaneous resources like picture-in-picture commentaries—the extraordinary growth of mobile browsing worldwide shows these exceptions are becoming rarer. As Silicon Valley legend and Greylock Partner in Residence Reid Hoffman noted, during the first internet bubble 15 years ago, investors played the 'Wrong Game'. They assumed distribution was inherently expensive and discoveries unlikely, so most quality content was expected to come from big old content houses with hefty advertising budgets. This was under oligopolistic conditions in distribution channels, where the chemistry of creative content was divorced from the mathematics of reaching consumers.

Emerging Markets and Local Content
YouTube's significant growth opportunity lies in markets like Russia, where mobile is often the sole method for consumers to access content, sometimes without easy access to payment methods or devices running the top operating systems. Google's limited accessibility poses a challenge, but efforts are underway to address this. Daniel Boldyrev of Google Russia illustrated how small channels

are making revenue through SMS-based payments and highlighted Google's initiatives to foster local content creation and support home-grown talent. Over the past five years, many Russians have transitioned from TV consumers to broadcast-quality TV producers, becoming a crucial segment of the YouTube user base.

Hyper-Local Strategy

Boldyrev also emphasized the importance of going 'hyper-local.' While Multi-Channel Networks (MCNs) producing and promoting content may not always succeed in these markets, focusing on local content that resonates with specific regional audiences can drive success. This approach underscores the value of tailoring content to fit the unique cultural and socio-economic contexts of different regions.

Competitive Landscape and Future Prospects

Google, completing its first 15 years and heading into what could be seen as its 'wisdom years,' faces mounting competition. However, in the online video space, it remains dominant, much like Microsoft 20 years ago, with a commanding market share exceeding 70%. Despite some disappointments, such as the phablet form factor, Google's strategies in international expansion continue to position it as a formidable player in the digital content landscape.

Conclusion

International expansion presents both challenges and opportunities for YouTube. By focusing on mobile accessibility, fostering local content, and adopting a hyper-local strategy, YouTube can continue to grow its user base and support content creators worldwide. As the digital landscape evolves, remaining adaptable and responsive to regional needs will be crucial for sustained success.

CHAPTER 9

Live Streaming

The Rise of Live Content
Live content is rapidly taking over streaming services, with film distributors and content owners investing heavily in this format. As in-house channels continue to attract views, we anticipate a significant increase in live broadcasts from TV networks and traditional cable channels in the coming year. This year alone, billions of dollars have been spent acquiring content for live broadcasts on platforms such as YouTube, Facebook, and Amazon.

Innovative Live Broadcasts
It's not just traditional content creators who are exploring live broadcasts. Innovative live content is emerging in various areas, including live sports coverage, concerts, news, and eSports. To meet the growing demand for live content, there will be an increase in both the production and acquisition of live content via the Internet. Traditional live satellite solutions will be supplemented—and in some cases, replaced—by more innovative and versatile solutions. For example, approaches like LiveU are redefining how live content is broadcast to both national and international audiences.

Advanced Technologies in Live Streaming
In sports venues, technologies like X-Reality are creating next-generation augmented reality (AR) content from live events, re-

defining audience experience metrics. Mobile app agencies such as Teleport and Switcher are also emerging, facilitating the production of high-quality live content for all types of events, including sports coverage. Both major content providers like TV networks and social media platforms, along with boutique mobile agencies, are tapping into the live streaming trend and providing viewers with the necessary innovative tools to comfortably consume live content.

The Impact of 5G on Live Streaming

In just five years, it is expected that most of the 5G content exchanged will be of the live variety. With the advent of 5G, societal content (including live) emissions will account for 60% of the total data exchanged across the Internet. Additionally, private content exchange will become the main mode of internal and private data sharing among the millions of smart devices used globally. Live events and other real-time "cultural" experiences, such as intimate family moments and pioneering moments, will contribute to the increase in private streaming and the broader wireless ecosystem.

Future Trends and Developments

To stay updated on these advancements and their repercussions, platforms like YouTube will continue to discuss and analyze future trends and updates. Before the widespread adoption of 5G, Netflix has been a major contributor to the increase in Internet activity, primarily due to traditionally recorded traffic. However, post-5G, other technology providers like AT&T and Amazon, large content owners such as traditional broadcasters and movie studios, and modern platforms like YouTube will become the main providers of streaming activities on high-bandwidth networks.

Conclusion

The landscape of live streaming is evolving rapidly, driven by technological advancements and increasing demand for live content. By leveraging new technologies and innovative solutions, content providers can meet this demand and enhance the viewing experience for audiences worldwide. As live streaming continues to grow, it will play a crucial role in shaping the future of digital content consumption.

CHAPTER 10

Virtual Reality and Augmented Reality

YouTube's Focus on New Tech Trends

YouTube's commitment to embracing new technological trends is evident in its ongoing efforts to enhance its platform for 360-degree videos. While it remains to be seen if it will be possible to watch these immersive videos on your phone without touching the screen, the company's ventures into virtual reality (VR) and augmented reality (AR) are noteworthy. By striving to capture a significant share of the VR market, YouTube is poised to accelerate the development and adoption of this technology.

Early Predictions and Developments

Patrick Llewellyn of 99designs predicted that 2017 would mark a significant year for VR, with companies like Facebook, Google, and other smaller firms pushing the technology forward. The debut of 360-degree virtual and augmented reality ads provided early evidence of this trend. Experts believe that YouTube will continue to leverage VR and AR to attract even more viewers to the platform.

The Popularity of 360-Degree and VR Videos

Although VR is not yet a widespread technology, 360-degree and VR videos have already become popular trends on YouTube. These immersive formats offer users unique experiences, allowing them to explore content in new and exciting ways. As YouTube's VR options advance, users will be able to share and watch VR content more comfortably and conveniently.

Enhancing User Experience with VR and AR

YouTube is likely to continue integrating VR and AR to enhance user experiences. This includes developing new YouTube remotes designed specifically for 360-degree video navigation and exploring touch-free interaction methods. Such innovations will make it easier for users to engage with VR content on their devices.

Potential Impact on the Market

By adopting VR and AR technologies, YouTube is not only improving its platform but also contributing to the broader adoption of these technologies. As more users become accustomed to VR and AR experiences on YouTube, demand for these technologies will grow, driving further innovation and development in the market.

Future Prospects

Looking ahead, YouTube's VR and AR offerings are expected to become more sophisticated. Users can anticipate more advanced features, better integration with other technologies, and increasingly immersive content. As VR and AR continue to evolve, YouTube will likely play a significant role in shaping the future of digital content consumption.

Conclusion

YouTube's foray into virtual reality and augmented reality is a testament to its commitment to innovation and enhancing user experiences. As the platform continues to develop new features and embrace emerging technologies, it will remain at the forefront of digital content creation and consumption. The future of VR and AR on YouTube holds exciting possibilities for both creators and viewers.

CHAPTER 11

Data Analytics and Personalization

The Power of Data in Viewer Hands

Data has become an essential tool for viewers, allowing them to adapt their viewing habits to a highly personalized, data-driven paradigm. As platforms and users develop an ecosystem together over time, audiences benefit from sophisticated personal recommendations. Initially, new YouTube viewers rely on the reputation of YouTubers to discover content, but as they generate more data online, channel subscriptions become almost redundant. This data helps the platform refine its recommendations, enhancing the viewing experience.

The Slot-Based Model and Feedback Loops

YouTube's capacity to understand its customers is significantly extended by its Slot-Based Model, a data-facilitated feedback loop that allows customers to express themselves in various ways. Viewers show their preferences through interactions on a YouTuber's channel landing page and features like the recommended list. This dynamic feedback system helps YouTube tailor content more accurately to individual tastes, creating a more engaging and personalized experience.

Algorithm-Driven Personalization

As discussed in the Algorithm section, YouTube is continually refining its data algorithms to present content more effectively to viewers. This ongoing development means more personalized changes to recommendations, subscription feeds, and even the home feed. Original release features are being added to the YouTube app, making the platform more interactive. As recommendations evolve, YouTubers may start developing shows that tackle niche problems not yet exploited, enhancing the content ecosystem.

Enhancing Viewer Engagement and Social Capital

YouTube's data-driven personalization can foster a deeper sense of engagement and social capital among YouTubers and their communities. With faster and cheaper data, viewers can build stronger connections with content creators, making the community aspect of YouTube even more critical.

Future of Data Analytics in YouTube

1. **Enhanced Personalization**: Continued improvements in data analytics will enable even more precise content recommendations, creating a more tailored viewing experience.
2. **Interactive Features**: New interactive features will be integrated into the platform, encouraging more active engagement from viewers.
3. **Community Building**: Data-driven insights will help creators build and maintain robust communities, enhancing viewer loyalty and interaction.
4. **Content Development**: Creators will leverage data to develop content that addresses specific audience needs, increasing relevance and engagement.

Conclusion

Data analytics and personalization are transforming the way viewers interact with YouTube. By leveraging sophisticated algorithms and feedback loops, YouTube can provide a highly personalized viewing experience that keeps audiences engaged. As the platform continues to evolve, the integration of data-driven features will play a crucial role in shaping the future of digital content consumption.

CHAPTER 12

Copyright and Content Regulation

The Role of Algorithms

We can draw a parallel between a YouTuber's relationship with their audience and a tweet that was read by merely 300 out of 1850 followers: both scenarios highlight frustration with algorithmic limitations. Algorithms often divide us from our latest creations, making it challenging to reach our entire audience. This issue is not unique to YouTube; it is prevalent across various social media platforms.

Potential Unionizing and System Updates

If Google doesn't update its Terms of Service (TOS) and strike system, we might see YouTubers unionizing or working within YouTube's system to seek better terms. Users could start dragging videos to the upload table from outside their accounts, potentially adding channel enhancements without significant extra work. YouTube mobile uploads were recently phased out, but full-length videos could still be managed through the Creator Studio to ensure upload satisfaction remains high.

Monetizing YouTube as a Legitimate Career

Imagine hundreds of YouTubers making a reasonable income, seen as a legitimate job by society. Young creators could start using simple devices like iPod touches to shoot and upload videos, rapidly increasing their subscriber base and income. However, challenges arise with the proliferation of platforms like Rumble and BitTorrent, which facilitate illegal content downloads. This leads to fewer copyrighted licensed songs or TV shows in video edits. Copyright law enforcement could become more stringent and impactful, presenting significant obstacles for creators.

Navigating Copyright Law

Copyright law is designed to protect creators' intellectual property, but it can also pose challenges for YouTubers. Unauthorized use of copyrighted material can result in strikes, demonetization, or removal of content. To navigate this complex landscape, creators need to:

1. **Understand Fair Use**: Familiarize themselves with the principles of fair use, which allow limited use of copyrighted material for purposes such as commentary, criticism, and education.
2. **Use Royalty-Free Content**: Utilize royalty-free music, images, and video clips to avoid copyright issues.
3. **Get Permissions**: Obtain permissions or licenses for copyrighted material when necessary.
4. **Content ID System**: Be aware of YouTube's Content ID system, which automatically detects copyrighted material and enforces copyright rules.

Community and Support

To support YouTubers, YouTube could offer more robust tools and resources for understanding and navigating copyright laws. This includes educational resources, support forums, and clear guidelines on fair use and licensing.

Conclusion

Copyright and content regulation present significant challenges for YouTubers. By understanding copyright laws, utilizing royalty-free content, and obtaining necessary permissions, creators can navigate these hurdles. As YouTube continues to evolve, ongoing support and updates to the TOS and strike system will be crucial in ensuring a fair and sustainable platform for all content creators.

CHAPTER 13

Advertising and Brand Partnerships

Revenue Challenges for Creators

In recent years, creators have faced significant challenges with the revenue generated from their YouTube channels. Many creators, who dedicate countless hours to their craft, have found that advertising revenue—a major income source—has become increasingly unstable. YouTube's AdSense program, once a reliable source of revenue for YouTubers, is no longer sufficient for maintaining channels with substantial followings. This decline in AdSense earnings has revealed several dangers for creators heavily reliant on this income source:

1. **Algorithmic Disruptions**: Events like the Adpocalypse have dramatically reduced monthly earnings for many creators. These disruptions can slash revenue without warning, making it difficult for creators to plan and sustain their operations.
2. **Staying Current**: To maintain views, subscribers, and revenue, creators must constantly stay on top of current content trends and platform features. This requires continuous adap-

tation and innovation, which can be exhausting and financially draining.
3. **Revenue Cuts**: YouTube takes a significant cut of advertising revenue before ads are run in conjunction with creator content. This reduces the overall earnings for creators, even if their videos are successful in attracting ad placements.

The New Landscape of Advertising and Brand Partnerships

Advertising and brand partnerships on YouTube have ushered in a new era of content creation and influencer marketing. Despite growing consumer skepticism towards advertising, major brands are increasingly compelled to engage with YouTube influencers, who hold considerable sway over audiences. Marketers who once focused on television are now pivoting their strategies to capitalize on the endless hours viewers spend watching digital video on TV screens and other devices.

Strategies for Successful Brand Partnerships

To navigate this new landscape, creators can employ several strategies to build successful brand partnerships:

1. **Authenticity and Trust**: Brands seek authentic connections with their target audiences. Creators who maintain genuine relationships with their viewers can provide value to brands through endorsements that feel natural and trustworthy.
2. **Targeted Collaborations**: Collaborating with brands that align with their niche and audience can enhance the relevance and effectiveness of partnerships. This ensures that promotional content resonates with viewers and drives engagement.
3. **Diversified Revenue Streams**: Beyond AdSense, creators should explore diversified revenue streams, including brand

partnerships, sponsorships, merchandise sales, and crowdfunding. This diversification reduces reliance on a single income source and provides financial stability.
4. **Content Quality and Consistency**: Maintaining high-quality, consistent content production is key to attracting and retaining brand partnerships. Brands look for creators who demonstrate professionalism and dedication to their craft.

The Impact on Marketing Strategies

Major brands are now playing the "YouTube game," focusing on digital video content to reach their audiences. This shift has significant implications for marketing strategies:

1. **Increased Digital Ad Spend**: As more viewers consume digital content, brands are allocating larger portions of their advertising budgets to online platforms like YouTube.
2. **Influencer Collaboration**: Brands are partnering with influencers to create engaging, relatable content that resonates with their target demographics. Influencers' ability to connect with audiences on a personal level makes them valuable partners in marketing campaigns.
3. **Content Diversification**: Brands are experimenting with various content formats, including sponsored videos, product placements, and interactive live streams, to capture viewers' attention and drive conversions.

Conclusion

The evolving landscape of advertising and brand partnerships on YouTube presents both challenges and opportunities for creators. By adapting to these changes, leveraging diversified revenue streams, and building authentic brand relationships, creators can navigate the complexities of this new era. As digital video consumption continues to rise, YouTube will remain a critical platform for marketing strategies, providing endless possibilities for creators and brands alike.

CHAPTER 14

Mobile Optimization

The Rise of Voice Searches

In early 2014, there were 2,123,481 voice searches across all platforms, including Siri, Google Now, and others. Experts predicted that the year would see a significant increase in spoken searches and verbal interactions with mobile devices. To prepare for future success on YouTube, businesses must consider verbal queries as a key component of their strategy. Relevancy is paramount in search, and the most relevant videos need to account for spoken questions and responses.

Enhancing Mobile Viewing Experiences

Users watching videos on YouTube need a seamless viewing experience—no squinting or half-screen displays. This requirement underscores the importance of high-quality video production and the increasing significance of good software in shooter gear over the next half-decade. Ensuring that videos are optimized for mobile devices is crucial, as viewers expect flawless playback on their smartphones and tablets.

Importance of Mobile Optimization

Mobile optimization is an essential aspect of the YouTube experience. In 2013, more than half of YouTube views were on mobile devices. By early 2014, YouTube's brand channels, customized for each brand, had an average of 130,000 views each, with 95% of those views coming from mobile devices. In the near future, mobile optimization will become a basic, expected element of the viewing experience, eliminating the need for users to check if their wireless devices are optimized for mobile content.

Strategies for Mobile Optimization

1. **Responsive Design**: Ensure that your videos and channel pages are optimized for various screen sizes and devices. Responsive design improves user experience across all devices.
2. **High-Quality Video Production**: Invest in high-quality video production to ensure that content is clear and visually appealing on mobile screens. This includes proper lighting, sound quality, and resolution.
3. **Mobile-Friendly Thumbnails**: Create engaging thumbnails that are easily viewable on small screens. Thumbnails play a crucial role in attracting mobile viewers to click on your videos.
4. **Optimized Metadata**: Use relevant keywords and descriptions that enhance searchability on mobile devices. Voice search optimization should also be considered.
5. **Interactive Elements**: Incorporate interactive elements such as cards and end screens to engage mobile viewers and encourage further interaction with your content.

Future of Mobile Optimization

As mobile technology continues to advance, mobile optimization will remain a critical factor in content creation and distribution. Voice searches and interactions will become more prevalent, requiring creators to adapt their strategies to accommodate these trends. High-quality mobile experiences will become the norm, and viewers will expect seamless playback, engaging content, and interactive features.

Conclusion

Mobile optimization is a vital aspect of succeeding on YouTube. By focusing on responsive design, high-quality video production, mobile-friendly thumbnails, optimized metadata, and interactive elements, creators can enhance the mobile viewing experience. As technology evolves, staying ahead of mobile trends will be essential for maintaining audience engagement and growing your channel.

CHAPTER 15

Social Media Integration

The Dominance of Large-Scale Influencers
OpenSlate estimates that the most viewed 2% of YouTube channels command 90% of all views and 94% of all subscribers across the platform. This dominance highlights the continued importance of large-scale influencers, despite the high number of dislikes on the recent YouTube Rewind. These influencers remain a core strength for YouTube. However, YouTube's direct competition with platforms like Facebook and Snapchat in the social media space must be approached with caution. The introduction of features like ephemerality and purpose-built audiences could quickly shift YouTube to a secondary position behind these platforms, until integrated features like Spotlight and stories gain traction.

Strengths and Weaknesses of Social Media Integration
Social media integration presents both strengths and weaknesses for YouTube. Despite the vast amount of data Google possesses on its users, leveraging this data for ad targeting on YouTube has not created a paradise of high ad rates. In fact, YouTube's advertising inventory is notoriously cheap. Additionally, YouTube's relatively insular nature on the user side may contribute to the superior ad rates seen on newer digital video platforms. For big brands and dedicated social network superfans, YouTube remains a hub for content that is

exclusively viewable within its ecosystem, requiring users to be both registered and logged in.

Enhancing Social Media Integration

To enhance social media integration, YouTube can explore several strategies:

1. **Cross-Platform Promotion**: Encourage creators to promote their YouTube content on other social media platforms like Facebook, Instagram, and Twitter. This helps increase visibility and drive traffic to their YouTube channels.
2. **Interactive Features**: Introduce interactive features that allow viewers to share YouTube content directly on their social media profiles. This can include share buttons, embedded video options, and social media widgets.
3. **Collaborative Campaigns**: Partner with social media influencers to create collaborative campaigns that span multiple platforms. This can help attract new audiences and increase engagement.
4. **User Engagement Tools**: Develop tools that allow users to engage with content across platforms. For example, integrating comments and likes from other social media platforms into YouTube's interface.
5. **Data-Driven Insights**: Utilize data analytics to gain insights into user behavior across platforms. This can help tailor content and advertising strategies to better meet audience needs.

The Role of Social Media Superfans

Superfans on social media play a crucial role in promoting content and driving engagement. These dedicated followers are more likely to share content, participate in discussions, and contribute to the overall success of a channel. By harnessing the power of superfans, YouTube creators can amplify their reach and build a loyal community.

Future of Social Media Integration

As YouTube continues to evolve, integrating with other social media platforms will become increasingly important. The platform must balance maintaining its unique ecosystem with leveraging the benefits of cross-platform promotion. By embracing social media integration, YouTube can enhance user engagement, increase ad revenue, and stay competitive in the ever-changing digital landscape.

Conclusion

Social media integration presents both opportunities and challenges for YouTube. By implementing strategies that promote cross-platform engagement and leveraging the power of social media superfans, YouTube can strengthen its position in the digital content space. As the platform continues to innovate, effective social media integration will be key to its ongoing success.

CHAPTER 16

Challenges and Opportunities Ahead

The Power of Social Networks

With the rise of social networks, online creators now have unprecedented distribution capabilities right at their fingertips. In the future, efforts to connect fans and creators through impromptu streams and conventional broadcasts will increase. Significant advancements in audio and visual technologies are on the horizon, including broadcasters utilizing 360-degree immersive video and integrating virtual and augmented reality objects. These innovations will merge with traditional media, transforming the viewing experience. Streamers who master 360-degree cameras, the Unreal Game Engine, or 3D character animations will gain a substantial advantage, pushing other creators to keep up. Distribution will become even more democratized, opening new pathways for creators to engage their audience.

The Current Golden Era for Online Creators

Right now is a golden time for online creators to build a career. However, success requires more than just ambition and posting regular videos. It demands technical knowledge, strong connections, and a bit of luck. Creators should start by developing a solid understanding of technology and utilizing tools from companies like Google, Amazon, or Microsoft to handle the heavy lifting. Leveraging machine learning platforms for automation, object identification, translation, and language processing can significantly lower content and operational costs. This allows creators to focus on their skill, creativity, and the core reason they are in this business: to produce high-quality content that resonates with their audience.

Key Opportunities for Creators

1. **Technological Advancements**: Embracing new technologies such as AI, VR, and AR can set creators apart from the competition. These tools can enhance content quality and viewer engagement.
2. **Machine Learning Platforms**: Leveraging machine learning for tasks like automation, object identification, and translation can streamline operations and reduce costs, enabling creators to focus on their creative strengths.
3. **Cross-Platform Integration**: Utilizing multiple social networks for distribution can help creators reach broader audiences and increase their content's visibility.
4. **Building Strong Networks**: Establishing connections with other creators, brands, and tech companies can open up new opportunities for collaboration and growth.

Overcoming Challenges

1. **Staying Updated**: Keeping up with rapid technological changes and evolving platform features is crucial. Continuous learning and adaptation are necessary for sustained success.
2. **Maintaining Authenticity**: As new technologies and trends emerge, maintaining authenticity and a genuine connection

with the audience is vital. Viewers value creators who remain true to their brand and values.
3. **Navigating Monetization**: Finding reliable revenue streams beyond traditional advertising, such as sponsorships, merchandise, and crowdfunding, can provide financial stability.

Conclusion

The future holds exciting opportunities for online creators. By embracing technological advancements, leveraging machine learning platforms, and building strong networks, creators can navigate the challenges and thrive in the evolving digital landscape. Success requires a blend of technical knowledge, creativity, and strategic partnerships, but the potential rewards make it a golden era for those willing to invest the effort.

Milton Keynes UK
Ingram Content Group UK Ltd.
UKHW021115271124
451585UK00017B/540